Canyon

By Eileen Cameron

Photographs by
Michael Collier

MIKAYA PRESS

Have you ever poured a pail of water over a sand castle and watched the rushing water make a long, deep cut through the sand?

Something like that happens when a river pours down from the mountains and makes a long, deep cut in the earth—a canyon.

Rock is harder than sand and it takes a long time, millions and millions of years, to make a canyon. No matter how long it takes, and no matter how deep and wide a canyon becomes, it all starts with tiny drops of water high up in the mountains. . . .

Water
    falls softly
        in cold snow crystals
            onto the mountaintop.

It melts into drops,
    and moves in trickles
        that slip through the crannies,

and driven by rain
slide into creeks
that tumble downhill,

and swelled by storms
foam over falls
and gush into brooks
that splash in the sun.

joined by swift side streams,
it cascades through the rocks
to rush in a river
that surges down channels,

and crashes through boulders,
   and with mallets of masses of pebbles
    and with sandy razors of silt
     it carves out the stone.

And through sun and rain,
    seasons and years,
        and stretching over centuries,
            as water wears down rock,
                wind weathers the stone,

and gravel scours the sides,
the river roars loudly as it runs,
and drives and hammers and pounds,
digging the river bed,
steepening the cliffs.

It sculpts,

forms,

shapes,

changing the land,
    chiseling through time,

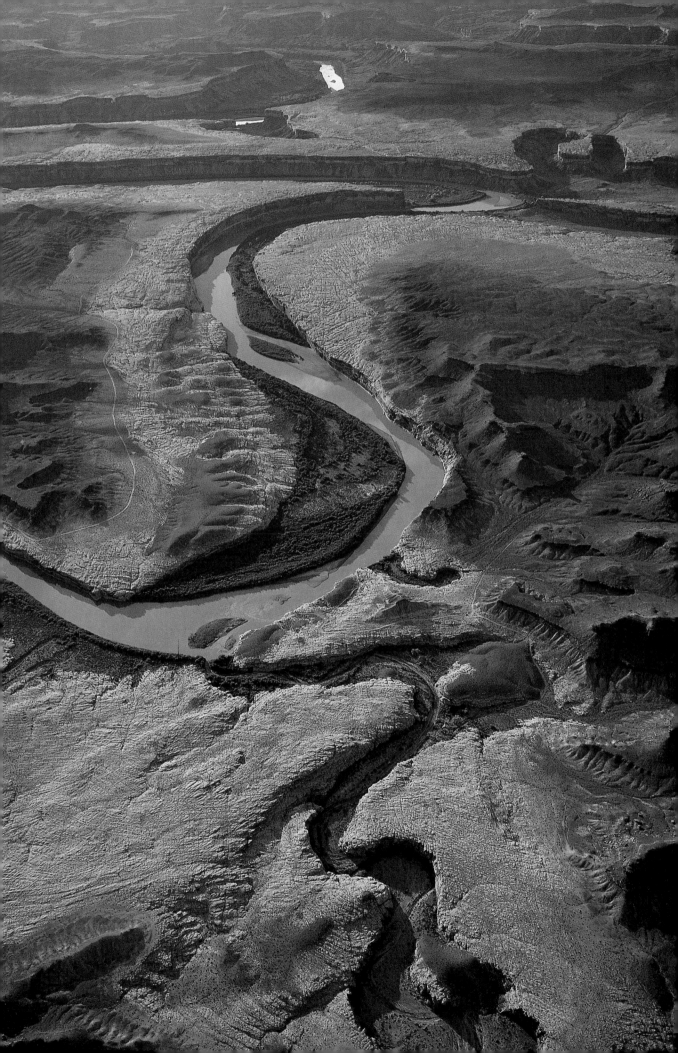

creating the canyon.

T he photographs in this book were taken along the Colorado River and the many streams, creeks and rivers that flow into it.

4. Thunder River

9. Antelope Canyon

5. Colorado River

10. Colorado River

1. La Plata Mountains

6. San Juan River

11. Bryce Canyon

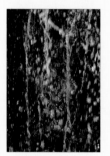
2. Stone Creek

7. Escalante Canyon

12. Green River

3. Curecanti Creek

8. Black Canyon of the Gunnison

13. Muddy River

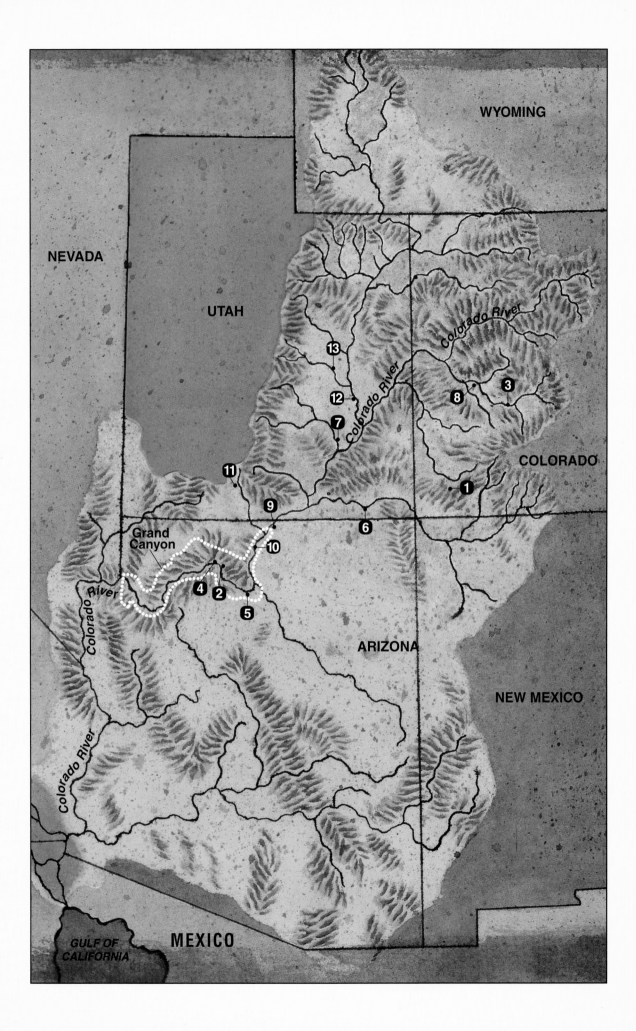

To my husband, Ed, and my sons, Colin and Ian,
my companions in hiking and admiring the grandeur
of our American West and its beautiful canyons.

Editor: Stuart Waldman
Design: Lesley Ehlers Design

Text Copyright © 2002 Eileen Cameron
Photographs Copyright © Michael Collier
Map Copyright © Alan Witschonke

Cataloging-in-Publication data available from the Library of Congress
ISBN 1-931414-03-3

Printed in China